FERVOR OF THE MINDS
Albert Einstein, Kurt Godel and Friends at Princeton(1942-2007)
A Novel
by Edward Nagel

"Out yonder there is this huge world which exists independently of us human beings, which stands before us like a great, eternal riddle, a cosmic religion at least partially accessible to our inspection and thinking through art and science."
Albert Einstein

"Godel was the only one of our colleagues who walked and talked on equal terms with Einstein."
Freeman Dyson

"The German man of science was a philosopher."
J.T. Mertz

"A Cretin says, "All Cretins are liars.""

Published in the United States by:
Quantum Literary Productions
2200 NW 9th Place
Gainesville, Florida 32605
e mail – writered33@outlook.com
www.quantumliterarygroup.com

First Printing – August 2015

Library of Congress Cataloging-in Publcation Data
Nagel, Edward - 1934
Fervor of the Minds – Albert Einstein, Kurt
Godel and Friends at Princeton (1942-2007)

ISBN- 10 – 1517126754

Printed in the United States
10 9 8 7 6 5 4 3 2 1

<u>DEDICATION</u>

For Harry Crews, (R.I.P), a kindred literary artist, and dear friend since our Halcyon Gainesville days when we studied under the late, great Southern novelist Andrew Lytle, then resident author at the University of Florida,for his support here for my novels and in New York City for my off Broadway plays.

"Engaging in philosophy is salutary, even when no positive results emerge. The color is brighter, that is, reality appears more clearly as such."

"We live in a world in which ninety-nine per cent of all beautiful things are destroyed in the bud."

Kurt Godel

"He who finds a thought that lets us penetrate even a little deeper into the eternal mystery of nature has been granted a great grace. He who, in addition, experiences the recognition, sympathy and help of the best minds of his time, has been given almost more happiness than a man can bear."

Albert Einstein's response to The British Royal Astronomical Society's award of the 1925 Gold Medal

"How do I see myself? With mixed feelings. I do not regard my achievements with calm satisfaction. I I am uncertain as to whether I was even on the right track. In me my contemporaries see both a heretic and a reactionary who has, so to speak, survived himself... The feeling of inadequacy comes from within. It cannot be otherwise when one has a critical mind and is honest, and mood and modesty keep us

Albert Einstein - 1949

CHAPTER ONE

CAMPUS OF THE INSTITUTE FOR ADVANCED STUDY – PRINCETON, NEW JERSEY – 1942

Kurt Godel, the thirty year old mathematician and logician, called by many the greatest since Aristotle, author of the Incompleteness Theorems, leaves his modest, compact home at 1451 Linden Lane. Dapperly dressed, sporting a new panama hat, he walks past, glances at and plastic flamingo standing on one foot in his yard. He strolls on under the mid -morning sun, arrives at the more manorial home of Albert Einstein at 115 Mercer Street and rings the bell. After a moment, Einstein, in his late sixties, appears, grinning, dressed in baggy pants held up by old world suspenders. They shake hands warmly, regarding each other with obvious fondness.

"So, Godel, dear friend, at last you are here among us. Would you care to come in, meet Maja, Margot and Helen, have coffee and strudel before we go to work, as it were, in the amusing, ceremonial backwater of the tiny, stilted demigods on the wonderful piece of earth called Princeton?"

"I think not Albert. I don't want to be late on my first day at the Institute."

"We don't keep time clocks around here, Kurt."

Einstein takes Godel's arm and they begin the first in a series of walks that would continue for fifteen years, prompting Einstein to say that while his work was all consuming, attempting to conflate general relativity to quantum mechanics, that he went to the Institute mainly for the privilege of walking and talking with Kurt Godel.

"Let's walk then, and talk. I'll set up a dinner for you and Adele to come over and meet your colleagues and wives. Bohr is here, just getting out of Copenhagen ahead of the Nazis. John Wheeler will come as well. They are here, with Oppenheimer, working on a project for the Government, all very hush - hush. Margot will cook one her gourmet sausage dinners and I'll play some of my favorite Mozart violin sonatas."

"We'd both love that... This walk to work with you is just what I need. I've got some projects spinning around in my head I want your views on."

Einstein stops to re-light his pipe.

"Perhaps extending your Incompleteness into new areas, such as models and recursion?"

"I did as much as I want to do on those subjects when I was here in '33. I want to look at your field equations for general relativity. ..You are involved in that hush- hush project as well?"

"Ironically, no Kurt. I gave them the very equation they are using to construct the Bomb yet I am deprived of helping them do it. Plus it was I who didn't think we'd see that Bomb in our lifetime, who reluctantly signed a letter to President Roosevelt to establish a program to develop a nuclear weapon in light of what Hitler and Speer are doing with Heisenberg and Hahn who have already effected nuclear fission in Berlin. I am ignored on that score though blowing up the world someday offends my pacifist views. That's why I am so pleased you've come so we can walk and talk like this and keep our dinosaur minds active, ward off atrophy and senility."

"Happy to oblige, dear friend."

Godel watches a car go by, and the driver who stares and grins at them.

"I suspect our walks will make for some interesting conversation among the natives, grist for their dinner party gossip mills."

"Yes indeed! Since arriving here in '33, I've reverted, much to Maja's and Margot's chagrin, to my Bohemian ways. I've become a sort of a village idiot. Oppenheimer thinks I've gone soft in the head for pressing my search for Unification. I've become a lonely old fellow, a kind of patriarchal figure who is known chiefly because he doesn't wear socks and is displayed on occasions as an oddity. Still, in my work, I am more fanatical than ever, entertaining the hope that I will solve the old problem of unity in the physical field. I feel it is right on the edge of the envelope, as I felt when I reached critical mass with relativity in '05 and '15. But it's much like being in an airship, cruising around, like but being unable to see clearly, how one can return to reality, that is, to earth."

"Seems absurd to leave you, of all people, out of the nuclear loop. Why for God's sake?"

"My politics, dear colleague, my history as a peace -nick. Hoover and his FBI minions think I'm a communist spy of some sort. Bohr and Wheeler are still here but lately I've noticed some of our best minds are being tapped on the shoulder to follow Oppenheimer to Los Alamos. But alas, no one taps me. We are in a race with Hitler for the Bomb and we better win it. Refusing my help hurts my pride but moreover keeps me from immersing myself in nuclear physics for my work on unification. And I might help Bohr who seems to be struggling with application of $E=mc2$. But not all is lost. The Navy has hired me at $25 a day to develop torpedo techniques , ironically what I did for Germany in 1920...How is Adele adjusting to Princeton?"

"Not as well as I am. I find Princeton much more congenial than Vienna now that the Nazis have overrun it with SS and street thugs. With the Jews being systematically driven out, the intellectual and creative core energy is gone from Vienna. You can feel the void, in theaters, coffee houses where I would meet with colleagues and friends. And Adele adored the night life, really thrived in that atmosphere. She would sing in the cabarets, which was, in addition to her being older that me, one of the reasons my family was so opposed to our marriage. If we had married earlier, we might have had a child now to care for...she gets so lonely when I'm away."

"I can relate to that. My family was dead-set against my marrying my marrying Mileva. We had Liseral out of wedlock. My father, on his deathbed gave us permission to marry but it was almost too late. That whole scene, Mileva and I growing apart, falling out of love when she failed to get her doctorate, losing the dream of us collaborating, our first born a bastard, was very painful and complicated my work on relativity. To atone for my treating her poorly when she withdrew from her studies, became depressed, I pledge and gave her the money from the Nobel."

"Did Liseral live on?"

"She survived a bout with scarlet fever but since she was illegitimate we had to put her up for adoption. We still get reports of her sightings, here and there. There seems to be a whole cottage industry, with the media, built on an obsession to find her."

"But you had two sons later?"

"Yes, Hans Albert in '04, who is doing very well as a civil engineer and then there's Edward, who sadly became mentally ill and is in Bellevue with Binswanger in Zurich. I last saw him in '33 after I became *persona non grata* in Berlin."

"Ironic that you are working for peace yet your work, I assume, is the basis for the ultimate weapon. And your work supporting Jewry has cost you. But certainly not here at Princeton."

"Not for a museum piece as me but yes, some antisemitic seeds have blown into the Princeton groves of academe in the form of the quotas for Jewish students and the PhD selection committees. I must admit that I didn't really know Jews until I came to the U.S., and saw the thousands who attended my lectures with the hate mongers, like that mob Hitler fires up at Nuremberg rallies."

"Their attack on the Jews will, I'm afraid, result in the ultimate horror but in the end that assault will be their downfall. They simply *cannot* be allowed to get the atomic bomb first. Attacking the Jews is, of course, designed by Hitler and his henchmen, to give the Germans a target to hate. Freud writes that the individual psyche's id, the irrational side craves destruction but is kept under control by the super ego. When Hitler arrives on the scene, after their defeat in WW1 and disastrous Weimar Republic, they quickly transfer their destructive drives to him, merge into a mob mentality and escape personal guilt. Hitler and his cohorts are mad dogs but masters of psychology hence a diabolical method to their madness."

"But how will it be their downfall?"

"They get immediate benefit and support from the hate factor and its transference into mob frenzy. But by driving out the Jewish scientists, like you, Bohr, Wigner and von Neumann, they are losing the very people who can get them the Bomb first. They are, no pun intended, cutting off their noses to spite their face i.e. the war effort. But you people...with champions as you on their side, will survive as they've done for centuries, almost like a nation, merely on their faith. It seems like a miracle to me and I will do all I can to help the cause here at Princeton and help and the other Jews."

"I wish to eliminate the need for miracles by pushing hard for a section of earth for the nation that has survived so long without one. Hitler's persecution, the diaspora of the Jews has had three good effects at least, has given us a specific enemy to battle and a hope for a Homeland and has brought *us* together so we may work on our projects. As an American Educator has said "Hitler shakes the trees and I gather the apples.""

The Institute is at hand.

"Ahh...here we are at the salt mines. Come, Wheeler and Bohr, two prime apples are in the building and will be delighted to see you."

"You and Bohr are on good terms after your rigorous debate at the Sixth Solvay Conference in Brussels? Has he convinced you that God does indeed pay dice with the Universe?"

Einstein laughs. "Hardly. But we still spare over quantum mechanics and are good friends. That debate was hardly acrimonious. Join us for lunch at the Commissary?"

"I don't eat out, Albert. You never know who's around to influence the food. I suspect there are a horde of Cretins put there conspiring to eliminate the scientists, the intelligentsia, to render us all idiots."

"But you must eat, dear man. You look thinner than when we last met...All right then, we'll see Bohr and Wheeler, set up something for dinner this weekend, with the ladies. Get your ducks in a row and at dinner we can return to our Halcyon days, relate how we came up with our inventions. You can go into your Incompleteness. We'll get Bohr to explain how he, with Heisenberg, shaped quantum mechanics. And I'll go back for a nostalgic look at my patent office days with my dear friend and sounding board Besso, and writing up those first papers. And as we talk, maybe some nugget of insight will break free for my unification quest."

"Not to discourage you Albert, but I feel, with our limited instruments that a merger of general relativity and quantum mechanics, a conflation of gravity and electromagnetism, is impossible. It's like, as you once did, trying to reconcile the paradox between Maxwell and Newton' ether. As the Americans say, "Apples and bananas.""

"Yes, but when I finally broke the paradox, in '05 on my trolley car in Bern, the golden symmetry of special relativity appeared. You are forgetting your Engels, Herr Godel. His simple dialectic. Everything is made up of opposing forces. The engine of scientific and artistic creativity...like Mozart's music...is thesis, antithesis, then a sublime synthesis. When the paradox serves its function, it dissolves into mist, leaving a paradigm shift and a simple, elegant picture that even a child can understand, as it did for me in '05 and '15. I feel that it is possible on Unification which is emeging in my mind as a scientific truth but not as a Western abstraction but as in the Greek sense, *alethes* – becoming unhidden, as a bright light in a chamber of uncertainty. Along with symmetry and pictures, metaphysics arises as well. I must create the marble of Unification out of the geometry of wood."

"As you are aware we are both Platonists, believe in the eternal truth in the form of math and physics. Meta-mathematics appears for me as well, when I apply my limit cases. I just ...*feel*..there is something still hidden in your general relativity field equations that even as great a discovery it was it is..how shall I put it without offending you..."

"Incomplete? No offense taken. But I honestly I don't know what you might do to extract something new from it. After all, I removed Newtonian cause and effect from particle physics, made time part of space, geometrized the Universe. What more do you think you can do with time?"

"I don't know...*yet* but with these walks and talks and hopefully the tutorials you will give me on general relativity, I'll think of something."

CHAPTER TWO

Seated around the long, oak dining table in Albert Einstein's home drinking coffee and after dinner liqueur were Niels and Margrethe Bohr, John Wheeler, Kurt and Adele Godel, Maja and Margot Einstein and Helen Dukas, Einstein's sister, step-daughter and secretary respectively. By the window at the end of the table Einstein plays the final part of a Mozart violin sonata and ends with a flourish. There is generous applause as Einstein bows.

Wheeler holds up his wine glass. "Bravo maestro. Wolfgang would have been pleased."

Einstein puts the violin in its case and sits.

"John, you are too kind. I'm hardly a maestro."

Maja rises and with the other women, begins to clear the table, then stares at her brother as he lights his precious pipe. "Remember dear brother, only one pipe per evening so make it last. Doctor's orders, "she adds, glancing at the others. "Anyone else for more coffee or liqueur?" Hearing no takers, she leads the other ladies into the kitchen, and after a moment, they return and sit back down.

Margrethe sips her coffee, says, "Dr. Einstein, you are quite the violinist. You take great comfort in the music of Mozart, Niels tells me."

"Yes, Margrethe, a comfort in the performance of the product of his amazing gift and yet, a step beyond that as well. To another realm, thru and beyond the notes. My love of Mozart goes back to my childhood. At thirteen, I discovered his violin sonatas. Beethoven *created* his music but Mozart's , music was pure genius, constructs seemingly always cruising the universe, harmonic constructs, waiting to be captured by the eye of the artist and reduced to notes. Mozart didn't just write the music music, especially the great late symphonies and operas, *happened* to him, in a flash of light and genius."

Godel chimes in. " As you plucked relativity theories from the cosmos?"

"Yes, precisely. I feel by the same method of discovery. Special and general relativity both *happened* to me in Bern. I am convinced that artistic and scientific universalities emerge from the same cosmic matrix. I identified with Mozart's ability to compose music even in very difficult and impoverished times, there at the end of his life. In '05 I was living in cramped quarters, dealing with money shortages, couldn't find work, having problems with Mileva. I sensed a disconnect out there in the physical world but I couldn't *see* it in my mind's eye. My ideas on space and time came from an esthetic of discontent in my choice between Newton and Maxwell. It seemed that asymmetries in 1905 physics concealed essential and elegant beauties in nature, existing theories lacked the architecture and secret unity found in Mozart and Bach, whose works were mathematical masterpieces. In those early days, for all my first papers, special and general relativity, whenever I felt I'd come to the end, to the cul-de-sac of paradox, an implacable tautology, I would take refuge in my music. And usually, together with my dear friend Besso in the Bern Patent Office, my sounding board, I'd resolve everything. It worked every time, except now here with Unification, the process seems to have stalled for some reason, denying me my simple thought picture."

Adele glances at her husband, as if remembering. "You went to your music professor as Kurtsy and I went, in the old days, to the cabarets and coffee houses in Vienna until the goose-step replaced the joy and rhythm of dancing in the streets."

Margrethe Bohr adds, "Or as Mozart had Costanza read him fantasies and fable when he was writing The Magic Flute, to keep him in the fantasy mood."

Niels Bohr takes a sip of liqueur. "Staying in the musical mode Kurt, I recently heard a great compliment for your Incompleteness Theorems, that they were a thorough ordering of several voices, mathematical, metaphysical counterpoint merging into harmonic chords never heard before. I would suggest that your work came out of the same cosmic matrix Albert speaks of. Nagel and Newman called your work an amazing intellectual symphony, combined with the paradoxical and the Alice in Wonderland world of Franz Kafka!"

Godel nods. "Yes, my Incompleteness was designed to give that affect, harmonic, a comment on, as was Albert's work has done, science as philosophy, a continuation of Kant whom we both loved so in our youth. And the Kafka comparison? I feel that is accurate, as my work, again by careful design and structuring, moving in then out and beyond The Liar's Paradox, roaming the magical and amusing Kafkan world, as K seeks the long ago canceled surveyor's job in The Castle. As Albert and as his beloved Mozart, who go beyond the notes and equations on the page, my math and logic reaches for realms beyond the numbers."

He looks at Einstein and grins. "Although I don't relate as strongly as you do to Mozart. I said I'd try harder if he would go with me to see Snow White and the Seven Dwarfs! Only fables present the world as it should be and as if it had meaning."

Einstein joins in the laughter around the table. "I met Kafka in Prague in 1911 at Bertha Fanta's place. He wasn't famous yet. The novelist Max Brod was there and later he sent me one of his novels. Kafka directed him after his death to burn all his novels and manuscripts. But Brod knew of Kafka's self canceling habits and didn't do it so we have all of those great works. Those were exciting times in Prague, the Jewish intellectuals meeting at Bertha's and the cafes."

Helen Dukas adds, "Bertha's daughter Joanna is here at Princeton, works in the map department at the library. She's most vivacious and a bit cheeky, sees Dr. Einstein almost every day, takes notes and keeps a journal on him so I wouldn't be surprised to see her work in book-form someday."

Maj rises and pours coffee all around. "It's so good and right to see everyone here, safe from the horror and madness that is on going in Europe. Albert is as loving as ever, still reads the Greek Classics to me at night. And we go most weekends with Mr. Bucky to Lake Saranac. I had to leave Italy because of Mussolini but my husband Paul stayed on in Geneva. Someday we hope to reunite, here in this wonderful place."

"We were all so fortunate to have gotten out of Germany with our lives and some property." Helen says. "So many didn't make it. And living here as Dr. Einstein's chief cook and bottle washer has been certainly adventurous. When we arrived in '33, I wondered if I'd landed in Hollywood by mistake. I went into a grocery store with a wad of twenty dollar bills that Elsa had given me, and that with my German accent, brought in minutes, the police with screaming sirens and the store was surrounded. I was questioned, the bills examined, then released, wondering if this would happen every time I went shopping! Living with Professor Einstein, I was accustomed to life turning into a circus but this was ridiculous."

There is general laughter around the table. Margot explains. "We found out later why they swooped in on Helen that way. The ransom money for the Lindbergh baby had been paid in twenties and the New Jersey merchants had been asked by the police to report any suspicious customers who might try to pass the ransom cash."

"Helen is right." Maja says. "We were fortunate to escaped when we did, when the Weimar Republic crashed, and Hitler made his move. They soon closed all the exists and captured all the Jews in their cars and took their jewels.

Einstein squints out the window at the gently lit community. "I had just spent two fine months with some good people at Cal Tech and was headed back to Germany, with some stops in Chicago and New York City planned, for some lectures and those speeches, urging the moral intervention of the world against Germany and Fascism, their attack on the Jews. The extent of their brutalities seemed to increase every passing day. I was shocked that the German people could not see what was coming."

"Hearing Papa's message in America, the Nazis took quick action," Margot offers. "Storm troopers raided his apartment five times in two days! But I had out- foxed them and they came up empty handed. I smuggled most of his important papers to the French Embassy in Berlin. When my husband Dimitri Marianoff called from out of town, I warned him not to come home but to leave the country at once and go to Paris, where I would meet him. Ilse and her husband, Rudolf Kayser had already escaped to Holland. The SS surrounded Papa's cherished vacation home in Caputh for arms and ammunition."

Einstein laughs. "They were the Nazi version of the Keystone Cops. They had been told by an informer I had allowed militant Communists to stockpile military stores on my property. They found nothing at first until they did an inch-by-inch search and finally found evidence of my arms stash – *a bread knife!* Instead of returning to Berlin I went to Antwerp and immediately resigned from the Prussian Academy. Planck was supportive but suggested I resign. They condemned my criticisms of Hitler charging I had violated the unwritten rule that a scientist must stay out of politics. I took umbrage at that stupidity, replied, where would we be had men as Giordano Bruno, Spinoza, Voltaire and Humbolt thought and behaved in such a fashion. Nazi outrages against the Jews were ongoing and more brutal. My property was siezed and I was declared an enemy of the State and a $5000 bounty was put on my head. I didn't know I was worth that much!"

Margot adds, "We ...Ilse, Helen and Walther Mayer, Papa's assistant, joined him and Mama, in a rather pleasant exile in Le Coq Sur on the Belgian coast. Meanwhile we had heard that in May, book burning rallies in Germany had begun with a vengeance. Tens of thousands of rabid students and beer hall goons, raided libraries and private homes, burned Papa's book's and essays along with those of Thomas Mann, Freud, Proust, Hemingway, H.G. Wells and Dos Passos. Propaganda Chief, Joseph Goebbels was there, amping up the crowds to frenzies, putting his official stamp on the burning as more than sixty bonfires blazed across Germany with Goebbels declaring, "Intellectualism is dead. The German soul can finally express it self. Papa said, "First they burn the books, then the people."

Niels Bohr stands and walks to the window...
"Such rubbish and insanity. I spoke with Planck
around that time and although not a Nazi he saw
the damage Hitler was doing to German science,
tried to to tell him that Jews could be good
Germans. Hitler showed how insane he really was,
telling Planck that had nothing against Jews but
that they were all Communists. When Planck
disputed that Hitler went into a rage, shouted that
people behind his back said that he had attacks of
nervous weakness but that he had nerves of steel,
ranting on until Planck left, convinced that there
was no hope of gaining a reprieve for his Jewish
colleagues, that they all had to get out of Germany
immediately."

"Kurt, on our first walk to work, pointed out
that irony," Einstein says. "Hitler is using the
hatred for Jewish scientists running them out of the
country thereby reducing their chances to get the
Bomb first. We got death threats in Belgium. Queen
Elizabeth, my dear friend, helped reduce the
tension, supplied detectives for our protection. But
safety was impossible even when the Secret Service
everywhere. Elsa was terribly worried so we left for
Switzerland to see Eduard for the last time, then to
England and here when Mr. Flexnor agreed to take
Walter with our little group here at Princeton."

Adele takes her husband's hand. "We experienced similar things in Vienna, when the Nazis marched in. Kurt had such a wonderful intellectual life in the '20s with his Vienna Circle, the coffee houses, the Akazienhof, where the white table tops were used to write out equations, and new ideas for art and philosophy. The night life and plays for me were wonderful. When the Nazis ran off the Jews, they cut the artistic heart and energy right out of the city. It got so bad when Kurt came back from Princeton in '33, we stayed as long as we could, got married in '38 and came here permanently in '40 due only to the kind influence of Mr. Flexnor who convinced them to let us go to America. The last straw, what propelled us out of Vienna was that one day on the street, strolling along, a street gang, thinking we were Jewish, attacked us."

Godel laughs. "And Adele, my faithful bodyguard, drove them off with her umbrella! One of the reasons I married dear Adele was for her gumption and her formidable presence. I tried hard to stay in Vienna, hoping against hope that some of the old day's ambiance would return, that the Nazis couldn't be that stupid to destroy the city's intelligentsia. But then again, that proves my main premise. that there are forces in the dark world who are conspiring to eliminate people like us."

Godel turns to Niels and Margrethe Bohr. "You both had some close calls as well on escaping from the Nazis did you not?"

"*Too* close," Margrethe answers. "Last September we were about gobbled up by those monsters. In '41 Denmark was under German occupation. When Heisenberg came to see Niels, on a mission to do something for Germany that we are yet to discover...we found that he, though he was working for Hitler on nuclear fission for war, at least helped us get out of Denmark."

Bohr agrees and adds. "He was one of our oldest friends. He discovered Uncertainty and later I added Complimentarity. He visited to see how we were doing."

Margrethe says, with irritation. "How could he be a friend Niels in '41? He was the enemy working for Hitler and Speer, to beat us to the Bomb. I never fully liked him...brash, too quick and eager. He came to rub our noses in it, that German controlled us Danes, that he and The Third Reich were superior. He *may* have been a friend when he arrived in '41 but after you and he went out to walk and talk, you were furious as what he wanted you to do, tell you how the US was progressing on the Bomb. That was the end of the famous friendship between Niels Bohr and Werner Heisenberg."

"I wasn't all that angry. More irritated than angry. We talked physics, not politics, on that walk and John, he said that what we did in '39 on the separation of natural uranium into U-238 and less than one percent of fissionable U-235 by fast neutrons, which would take forever to extract for bomb use, and he said that was wrong."

"But you were right," Wheeler asserts. "And later based on your insight from that walk on campus in '39, your "Bohr moment" as you predicted 238 would absorb the fast neutrons and would be transformed by them into a new element, Neptunium, which would decay into Plutonium, at least as fissionable as the 235 we couldn't practically separate. But was that comment what made you so angry at him?"

"No, not really. That was such a tense time for us all, it has clouded my memory. He asked me I'm sure if a physicist had the moral right to work on the practical exploitation of atomic energy.

"And a bomb went off in your head," Margrethe says. "You jumped to the conclusion that he meant the atomic bomb and that he had come to pick your brain as to what Fermi was doing in Chicago and what you, Oppenheimer, Feynman and the others were doing in Los Alamos, in obtaining fission and building a functional Bomb."

"Perhaps so. But he said he wasn't referring to an atomic bomb but only to build a reactor to produce power for peaceful purposes. He was torn between his alliance to his beloved homeland and trying to get the Bomb for Hitler, knowing that possessing the ultimate weapon would let crazy Hitler to destroy the civilized world. I gathered from our talk, that he may have by mistaken math in the way he computed the amount of U-235 needed to make a nuclear weapon or he computed it to be so high that Speer who controlled the money, whom Hitler trusted implicitly, might not give him the funds to continue or turn down the project. He knew of Speer's timetable of two years to win the war. And if the Allies air raids and commando attacks had not destroyed his ability to make heavy water, we all would be under Nazi rule now, held prisoners here, that is if Princeton even existed."

"Off the record," Wheeler says, "we were behind the nuclear fission race. Hahn and Strassmann, after Chadwick discovered the neutron which could split the uranium nucleus, smashed the atom in Berlin in '38. Niels brought us that wonderful news from Denmark which alarmed us all. In Chicago we received a cable from Denmark from Houtermans in Switzerland in connection with

German work on uranium fission, urging us to get going for fear that Hitler would get the Bomb first. That was our wake-up call."

Einstein remembers. "I was on holiday in July at a summer resort on Long Island when Wigner brought me the news of Germany's fission success. I wasn't totally surprised as I was so involved with other work but had thought that while it was possible, this weapon could not be invented in our lifetime. But they did it. When Wigner explained the process I immediately understood it...then realized in flash that it had gone beyond science to become a political decision. From that moment when I was asked to write the President and urge we "get going" as Wigner said, I was and remain torn, sort as Heisenberg was, as Niels said, between my peace efforts and the need for us to jump in with both feet into the nuclear race to avoid that takeover by the Nazi abomination. After our letter to Roosevelt they finally set up a committee to study the project. In November Columbia University got a grant and hired Szilard and Fermi to produce a chain reaction with uranium and well, as they say, that universal, anonymous "they", the rest is history."

Bohr adds, "Yes, we got started but the FBI almost torpedoed the project or at least harder to begin. We had asked Albert to calculate the best way to purify the uranium but they kept vital data from him as they thought he was a Communist and without that we got going down the wrong road. The splitting of the uranium atom had released 200 million electron volts of energy apparently out of nowhere. But we were puzzled – where did the missing mass go and where did this mysteriously come from? Meitner finally realized that Albert's equation- $E=mc2d$ held the key to this puzzle. If one takes the missing mass and multiplied it by $c2d$, then one found 200 million volts, precisely according to Albert's equation. When I discovered this in a moment of epiphany, I was stunned. What fools we all had been!"

Margot says, "We have the impression that now we are in a tight race with Germany for the Bomb. But from Margrethe tells us about seeing Heisenberg last year, and what Dr. Bohr said he learned later about Speer and Heisenberg, is that really true?"

Wheeler responds. "We feel that keeping that view as for the general public consumption active here and in Los Alamos will maintain the required sense of urgency for the workers and scientists. But that might not be the actual case."

"Heisenberg's visit was as I mentioned was disturbing and puzzling but one thing is for sure, we are grateful for his help in getting Niels out of Denmark. When they were about to move on us, something sublime happened," Margarethe says.

Bohr agrees. "It *was* sublime. And Heisenberg had a sub rosa hand in it. That night last summer when I escaped across the pond in the fishing boat , the freighters arrived from Germany that Wednesday to pick up the Jews which numbered about 8000 in Denmark to be arrested and crammed into their holds. That next Friday evening, at the start of the Sabbath when the SS began their roundup, there was scarcely a Jew to be found. They all had been hidden in churches and hospitals, in people's homes and cottages. How was that possible? Because we'd been tipped off by someone in the German Embassy!"

Margrethe adds, "Heisenberg told us it was George Duckworth, the shipping specialist, one of his men had pulled it off. He was a remarkable informant. He told us the day before the ships had arrived, the very day that Hitler ussued the order. He gave us the exact time the SS would move. The Resistance got the Jews out of their hiding places and smuggled them across the Sound."

"It was an amazing bit of Jungian synchronicity,"Bohr says. "For a few of us in a fishing boat to get past the German patrol boats was intriguing indeed. But for a whole fleet to get past, with those eight thousand Jews on board was like the parting of the Red Sea! There were no German patrol boats that night, the whole squad had suddenly been declared un-seaworthy. Heisenberg didn't take all the credit for my escape although he was the prime mover in the effort. He said it was his man Duckworth again who also went to Stockholm and asked the Swedish government to accept everyone. It's acts as this that restores one's faith in the goodness of humanity, even as Hitler as an aberration rampages through Europe."

Helen nods. "Heroic work. Too bad the same thing could not have been done for the doomed Jews around Europe as they were taken off to God only knows where. Dr. Bohr, that was a very close call."

Bohr laughs. "But I wasn't out of the woods yet. A British plane flew me from Denmark to Britain but a badly fitting oxygen mask almost killed me. An alert pilot saw I was passing out for lack of oxygen and quickly dropped to sea level and I recovered. The way we all got here is proof I feel that some God- like Wizard behind the curtain is doing all he...or she...can do to defeat the executioner."

Maja stands. "Yes, we are so blessed to be here

in this safe haven. It was splendid having you all dine with us. I must excuse myself for the night. Margot, can you come with me and maybe read me the book that Albert started last night? I'm sure dear brother and his friends have more to discuss. And perhaps Helen can take it all down and transcribe it so I and maybe others in years to come. can read it for a look at this part of the world and what you scientists are doing to preserve our way of life."

The men stand as Maja and Margot leave, then sit back down, except Wheeler who raises his glass for a toast. "It just occurred to me that here gathered we have an auspicious group of innovative authors of the most fundamental scientific principles of the 20th Century. With Albert we have Special and General Relativity, Kurt with his Incompleteness and Niels, with Heisenberg ironically across the pond working the other side of the street, the Copenhagen Interpretation of quantum mechanics, a confluence of the Zeitgeist – the spirit of the age. To you gentlemen and the ladies who provided love and support while you brought forth these brilliant insights on the workings of nature, I raise my glass in a toast of thanks on behalf of mankind both past and future.

So often in history of the arts and science the ladies, wives, the women "behind the men" so to speak who provide the intuition of the distaff side, the glimpses and whispers of and from the Moon, are rarely given the credit they deserve."

All at the table raises their glasses and join in the toast as Wheeler sits and continues.

"I think it's intriguing that each of you reached an ontological conclusion about reality through the use of an epistemic principle concerning knowledge. The dance or dialectic of knowledge and reality is the dominant theme of the Twentieth Century. But, paradoxically, though all of your works opened areas of knowledge in arts, math and science are of immense value, they all established profound but somewhat disturbing limitations."

Helen rises, goes to the desk and gets a tape recorder, turns it on. "If it's all right I'd like to record this session subject to your reading it later for any classified information which I will delete, then place in Dr. Einstein's archives, so as Maja suggests, it can be published for others to read of this intriguing time in history.... Dr. Wheeler can you discuss in more detail why it is paradoxical?"

"Limitations and openness?" Wheeler muses. "Albert's relativity set a limit – the speed of light – to the flow of any information bearing signal. By defining time in terms of its clock measurements he set a limit to time itself. Heisenberg's Uncertainty principle, coupled with Neil's Complementarity their quantum mechanics, set a limit on our simultaneous knowledge of the positions and momentum of the fundamental particles of matter. And with Kurt's Incompleteness..."

Godel interjects. "This limitation was not just on what we can know. For Heisenberg it signified a limit to reality. For me and I think for Albert , we saw quantum mechanics as a mere heuristic principle, as opposed to deducing the nature of deep reality from limitations on what can be known. It became, for Heisenberg ...and no offense, for you Niels...a kind of religion."

"No offense taken,Kurt," Bohr says. "We didn't want our findings to be blown up into that as we were both shocked by its implications if taken as a law of nature. We searched high and low, had our grad students drill us on its authenticity and depth of influence. And of course there was Albert and myself dueling in thought pictures at the Solvay Conference, where my noble adversary said quantum physics was not the true Jacob. And that God does not play dice with the Universe."

Wheeler continues. "And on the limitation from Kurt's Incompleteness, called recently at Harvard as the most significant mathematical truth of the Century, set a permanent limit to our knowledge of the basic truths of mathematics. A *total* set of mathematical truths will never be captured by any finite set of axioms. If we can grasp the whole truth, our minds are not machines, not computers."

Adele turns to Einstein. "Kurtsy tells me that you and he had similar backgrounds in your youths. He started studying physics but later switched to mathematics and logic. You started in math, held in high regard your text of Euclid, then switched to physics?"

"Yes, I began in mathematics but thought it only a tool to access higher truths in physics, that by itself it could never come up with anything fundamental until I saw what Kurt did with Incompleteness. He went beyond the numbers outside the digits, went outside the Liar's Paradox to realms of meta-mathematics."

Adele adds. "And you both, I understand at a young age began the study of Kant?"

"I discovered Kant at thirteen, that tiny eighteenth century philosopher who proposed some bizarre ides, such as we have been or will have been inhabited by some form of life. As a stand-up comic, he peppered his lectures with asides, shot down the pompous philosophers. But his comments on time and space – that they are not products of experience but concoctions of our minds which which clothe our sense perceptions, reinforced my already aroused interest in the subject. Kant denied the existence of a personal God which I agreed with as the Judea/Christian religions are based on fear of being damned to an absurd concept of hell after an further absurd method of determining who goes to heaven or to hell. Also, Spinoza was a favorite of mine."

Godel adds with some irritation. "Which is why I feel, and believe Albert thinks so as well, from what we have gathered from our talks, that the public erroneously thinks that we have, in our work, put man in the center of the Universe, have done away with Newtonian determinism, without any objective knowledge, as opposed to the subjective view of quantum mechanics. Which is totally false."

Einstein nods. "John is right when he says that our work, Kurt's and mine, are on two prongs of the Zeitgeist. The quantum mechanics has most famously taken root as the third prong."

Einstein looked away. "The Unification paradigm is not emerging as my past concepts had, from the simple picture, then the breaking down of the paradox and those flashes of insight, "happened to me". The fervor of intentionality for the discovery of the theory of everything as some people call it, is there all right. But perhaps my advanced age has reduced my power of inductive and intuitive reasoning. But I think not. My mind seems as sharp as ever, as it was in my Halcyon days of our Olympian Society, but I may have intentionally refused to deeply study quantum mechanics because of my dislike in how it has been used by some to totally demolish, across the board of modern physics, determinism, to catch on as most around me have, to the deep richness of the subatomic world...the features of spin, mass, charge and quantum numbers and that this lack of knowledge is preventing me from discovering the key to unification. I do have at least the working picture or metaphor in my mind to transmute the wood of quantum mechanics, on the lathe of general relativity, hopefully rendering the marble of unified theory--much as I feel the alchemist transforms, at least in his dreams, base metal into gold. I just don't know enough about the complex properties of nuclear force."

Godel looks at Bohr. "Which is why it seems a shame and idiotic that the stupid paranoid FBI are preventing Albert from working on the Bomb considering that time is of the essence. And especially when your E=mc2d is the engine and core of that effort and by letting your genius directly influence the progress of the search at Los Alamos...and as a by product you might get that flash of light reveals the picture of Unification."

Margrethe adds, looks at Einstein. "Perhaps Dr. Einstein if you do see it fully formed, ready for you to do the math, we will see finally the light of your Old One's eye and mind. From those first days, when you gave up on a personal God, did anything happen to go beyond the personal religious guest, a new kind of acceptable religion?"

Einstein stands, walks over to the large open picture-window and stares up into a moonless sky, filled to bursting with stars and galaxies and seems to drift into another realm of thought, as if he is zooming back to 1905 and those miracle set of papers in wrote while in the Bern's Patent Office.

" Religion Margrethe? When I did away with the personal God, out went the usual anthropomorphic concept, where people pray and really believe that this man-made God, actually hears their prayers for personal favors and *responds*. and the ridiculous belief that at the end of the world, the dead will arise fully formed and those who have been saved will go to Heaven and the rest doomed to a roasting by a flame that burns forever. From when I first

started to read greatly, then during my time with
Mileva in Bern, while in the Patent Office with my
friends in the idyllic time of our Olympia Academy,
studied great art and science with Solivine and dear
Michele Besso at the office, my confidant and
sounding board. Some think that my time at the
Patent Office was a waste of time but it gave me
living money and a base for the theoretical work I
did later, when I had take those ideas offered for a
patent, forced to understand and see the shape of
structure, testing them against existing standards,
even making adjustments and sending them back to
the inventors for fine tuning gave me valuable
exposure to the trial and error method, coupled
with the thinking in electromagnetism as I worked
in my father's electric shop, gave me further
foundations, what most probably pushed me in the
direction of Maxwell's electric forces in modern
times and taking Newtonian physics out of my
pictures of relativity. But from that time to now, as
I struggle with Unification, it'd become clear that
the religious paradise of my youth, which was lost,
was a first attempt to free myself from the chains
of merely personal and primitive feelings, which
resulted in our modern theology as a religion that is
an opaque ideology institutionalizing alienation, a
defense against guilt and anxiety."

Einstein again, as if being drawn into the whirling cosmos arced out over the night sky, in a voice resonant with fine and trans-formative ideas, describes what he sees in terms of a cosmic religion. "Out yonder there is this huge world which exists independently of us human beings, which stands before us like a great, eternal riddle, partially accessible through art and science, to our inspection. The road to this paradise was not as comfortable and alluring as my childhood religious beliefs but it has proved itself as trustworthy and I have never regretted having chosen it."

Godel extrapolates. " That is why I think Albert and I are such compatible walking and talking companions. We both are committed to the task of reaching out, to describe that "out yonder" world. Only my domain is mathematics and logic dealing with objective reality, also known as mathematical Platonism, in honor of the ancient philosopher Plato whose metaphysics rejected the Sophists battle cry, "Man is the measure of all things." Philosophy without ontology is an illusion. Physics and mathematics without philosophy is reduced to engineering."

Bohr says, "You were a rebel in your Vienna Circle, in the positivists with Wittgenstein as their mentor. How did you survive that Kurt?"

Godel grins. "By adhering to Wittgenstein's famous dictat. "What one cannot speak thereof one must be silent." I was a kind of Platonic spy, sat silent, only nodding when I agreed to their postulates. If I had announced my anti-positivist leanings I would have been most certainly run out of the Circle.

Adele asks, "Was your Olympian Academy similar to Kurtsy's Vienna Circle Professor?"

Einstein disengages from the probe of his out yonder territory, blinks himself back into the present. "Adele, his Circle had a much higher level of academia, more deep scholars than our Olympia group. But they both, in terms of giving us a base from which to develop our theories, was similar. Kurt has told me that he was a kind of platonic spy in the house of the positivists,didn't add much but was building the base for his groundbreaking logic and math as being one side of the dialectic, the antithesis to his sublime premise for incompleteness. My special relativity came out of the same kind of dialectic, on the wave of our youthful, scientific exuberance reaching out for Olympia."

Helen asks, " What can you tell us of those years and how your relativity evolved into what is today, and how you wrote those five great papers in '05?"

Einstein looks fondly out the window, grins. "I felt the new ideas in art and science came from the mind of that God whom I called The Old One, who permeates out yonder, beyond the parameters of man-made theology. I had an uncomfortable feeling which arose from the increasing difficulty in explaining the deficiencies of Newtonian mechanics in light of Maxwell's combining the magnetic field to create an electrical field. His work produced a cyclical pattern which would create a never ending wave and the speed of this wave was the speed of light, unearthing its deepest secrets. A critical mass was building between the theories of Newton and Maxell as they were mutually exclusive and convinced me the world was ready for a revolution. But was I the one to lead it even when I was not totally sure Newtonian mechanics was not the future of particle physics, that it was incompatible with Maxwell's theories, and that one of the pillars of science must fall."

Adele says, "Can you recall the time you had the breakthrough that gave you special relativity? You have said that you had a surge of insight, a picture of the theory. Kurtsy complains that he has never has these kind of pictures to start his creative process. When he gets new ideas, he starts writing down symbols."

"In 1905 I had been haunted since childhood of the picture of what would happen if you raced with a light beam. I had read with fervor, David Hume's his paper , A Treatise of Human Nature, along with my study of Mach and Faraday, his experiments with the relativity between a coil and a magnet. It is possible that without these philosophical studies I would have not invented special relativity. I was intrigued by Maxwell's criticism of Newton's theory, said that the speed of light remained the same no matter how you measured it. I puzzled over this because in the Newtonian common sense world you can always catch a speeding object just by increasing your speed. The picture I used was a motorcycle police officer chasing a speeding motorist. He will always catch them. But replace a motorist with a beam of light and when we observed the chase, if the officer can go the speed of light, he will, from a distance appear right beside it. But when you question the officer about his chase, he will tell you that no matter how fast he went, the light beam traveled away from him at the speed of light as if he were standing still.!"

"And that was your core picture for special relativity, that began the process of discovery?"

Einstein nods. "Yes! This was the central, seemingly implacable paradox. How was it possible, I asked myself, for two people to see the same thing and come up with different conclusions? The germ of special relativity was already in that paradox. I went to Besso with that conundrum. We talked about every aspect of the problem. It seemed that Newton's idea of absolute space Maxwell's constant of the speed of light were elements of it. I became totally exhausted and told Besso that I was defeated and would give up the entire quest. I had failed and I was resigned to failure but with my thoughts still churning but stuck in the tautology."

Bohr adds. "But out of those churning thoughts, kind of intellectual chaos came your breakthrough right Albert? That process is familiar to us physicists: Heisenberg and I did the same thing . We pushed the paradox to its breaking point. Heisenberg, in Heligoland, walked around Faelled Park, saw himself up in the mountains, looking down with a telescope, saw himself in fleeting glimpses, moving in and out of streets and buildings, like the electron that moves around in the cloud chamber, not a continuous track but a series of collisions between the passing electron and molecules of water vapor."

"I left Besso, vowing to give up my search.

But as I sat on the street car going home and passed the huge Bern Clock Tower, I imagined what would happen if the street car raced away from the clock at the speed of light. I realized that the clock would appear stopped since the light could not catch up to me but my own pocket watch would keep the correct time. Then it hit me in that flash of light, from an unhidden state of uncertainty, a storm broke loose in my mind. The answer was simple and elegant: *time can beat at different rates depending how fast you are moving!* One second on earth was not the same as one second on the moon. Even though I could not afford a decent clock for our home, I placed a clock at every point in the Universe. I had, it seemed, tapped into the Old One's thoughts! That solution which came to me so suddenly with the thought that our concepts and laws of space and time can only claim validity insofar as they stand in clear relation to our experiences. By a revision of the concept of simultaneity into a more malleable form, I thus arrived at special relativity. The next morning I went back to Besso and blurted out, 'Thank you! I've completely solved the problem. Time cannot be absolutely defined and must be a function of space as well. There is an inseparable relation between time and signal velocity.'"

Margrethe asks, "That was it then? Did you need anything else?"

"To complete the theory, I incorporated the Lorentz-Fitzgerald contradiction, except that it was space itself that was contracted and not the atoms. I owe more to Maxwell than to anyone and thereby dispensed entirely with the aether theory. For the next six weeks, I furiously worked out every mathematical detail, then gave it to Mileva to check my math and went to bed for two weeks. I called the paper, On The Electrodynamics of Moving Bodies, scribbled on thirty one pages, which were published in '05 with two other papers."

Wheeler remarks, "Yes indeed, the famous Volume 17 and one of the most remarkable volumes in the whole of scientific literature, that one page describing what Niels and I work on, back at the shop – E=mc2d!"

"That came a bit later John, in a small paper, almost a footnote to Volume 17 which set the core idea that the laws of physics are the same in all inertial frames and the speed of light is a constant in all inertial frames."

Einstein lights his pipe. "Looking back, it's amazing to me now how close others had come to getting relativity going as I did but they dared not break out of the Newtonian lock-box of subject/object language. Lorentz and Fitzgerald obtained the same contradiction between Newton and Maxwell but thought it was a mechanical deformation of the atoms, rather than a subtle transformation of space and time itself. Poincare understood the speed of light must be constant but refused to abandon the Newtonian framework of the aether and thought these distortions were strictly a phenomenon of electricity and magnetism. For $E=mc2d$ I reasoned from special relativity, if meter sticks and clocks became distorted the faster you moved, then everything you can measure with meter sticks and clocks must also change, including matter and energy. Which meant that energy of motion was somehow being transformed into increasing the mass of the object. Thus matter and energy are interchangeable!"

"An amazing journey to genius Dr. Einstein," Margrethe adds. "Did your special relativity catch on quickly in the world of physics?"

"It was, dear lady, to my dismay ignored until '06 when the first glimmer of response came in the form of a letter from Max Planck, who saw that my work elevated the speed of light to a constant. His constant Planck time demarcated the classical world from the subatomic realm of quantum mechanics. We were shielded from the strange properties of atoms because of the smallness of of Planck's constant. He saw that by my raising the speed of light to a constant we were shielded from the equally bizarre world of cosmic physics by the huge value of the speed of light."

Hellen grins. "Dr. Einstein tells the amusing story of the reaction of one of his college professors Hermann Minkowski to his first theory of relativity."

"Yes, that *was* funny," Einstein agrees. "In college, to my later dismay, I ignored subjects that bored me, prompting Minkowski to call me a lazy dog for cutting classes. When special relativity first came out, he rewrote my equations, which I initially resented, felt the elegant picture had been ruined by mathematics, considered his work a superfluous erudition. But I saw that his work had enhanced, rather than distorted my picture. His rewrite had revealed this beautiful four dimensional structure forever linking time and space into a four dimensional fabric so that time and space vanished into the merest of shadows. I use his work even today as I work on Unification, the rule that the more beautiful, simple and elegant an equation is, the more phenomena it can explain in the shortest amount of space."

Adele asks, "And after that your reputation began to rise as a physicist?"

"Yes. In '08 my work was recognized as a breakthrough and on the wave of this notoriety, I applied for the post of professor at the University of Zurich where I had been treated with contempt for not knowing how to get along with important people. My competition for the job was Friedrich Adler, who had better politics, also a Jew. But he graciously gave it up in favor of me. So I returned to Zurich as a professor and not as an unemployed physicist and misfit, rebellious student."

Bohr adds, "And the irony continued! Your first lecture there gave Heisenberg and I the base premise for Uncertainty and Complimentarity. You posited for the first time the concept of the duality in physics, the proposition that light can have dual properties, can be either a wave or a particle depending on how you measure it. Some now call you the grandfather of quantum mechanics. Some might say, seeing how we argued over quantum mechanics at the Solvay Conference, a kind of Pandora's Box for you Albert?"

"Yes indeed it did, "Einstein laughs. "What do they say, what goes around comes around. The general understanding even among physicists who should know better, is that I despise particle physics. But that is so only if as they did after you all brought the theory to light and to Solvay, they claimed it *entirely displaced determinism!*. I was shocked when my mentor Max Born agreed. But it is not the true Jacob though it is useful to understand the dynamics of particle physics...As in that lecture I gave in Salzberg where I brought E=mc2d to the world and now it is coming at us in the form of the ultimate weapon. But soon after that I began to lose interest in special relativity. I had given birth to it but then as any loving parent, I immediately realized its potential faults and tried to correct them."

"And that dissatisfaction led you to start on general relativity?"

"Yes, I began then as the success of relativity had boosted my scientific stature and helped me get the Zurich appointment. But it was time to move on and chase the illusive power and mystery of gravity."

Bohr says, "There seemed to be at he turn of the century as you mentioned Albert a dissatisfaction with the physics of the time a feeling that there was more to the natural world than what Newton coul d tell us. In colleges, before Madam Curie came up with her radioactive discoveries and you did relativity and we did quantum physics, professors were telling physics graduate students to get into another field as Newton's aether and billiard ball physics were the end of the line, that the future was laid out no mystery in the Universe. The need for something new was brewing in the arts and psychology as well as the poet Ezra Pound has pronounced his famous dictat to the European writers and artitsts, "Make it new!" Out of that time came the Cubists – Picasso and Kandinski, Kafka, Proust, Joyce, Virginia Woolf and Luigi Pirandello, along with Freud in Vienna and Jung in Zurich. And new theater in Russia with Chekhov and Stanislavski."

Einstein grins. "I met Pirandello in Paris, where I saw his great modern play SIX CHARACTERS IN SEARCH OF AN AUTHOR, and found him, with his relativistic, quantum solipsist perception in his drama, a kindred spirit. While in Zurich, my assistant Hopf, had an interest in psychoanalysis, which led him to Carl Jung. I understand Wolfgang Pauli was a friend of Jung as well and went out with his secretary. When I expressed a curiosity about the man, he arranged a meeting for me with Jung. This meeting led to more. We were both out of our depths, in strange waters and with no lifeguard around. But we managed to keep afloat and in touch. Jung told me that he was no mathematician or physicist, but he was seeing a correlation, which he believed is the key to discovering new concepts, between my work on relativity and the quantum wave function field, was coming closer to the nature of the psyche, the World Soul, or Anima Mundi as he called it, the action of synchronicity which is out side physics, but exerts a powerful influence in how we move in the world, as if players moved by a master playwright as Pirandello."

"Did you ever meet Freud, Doctor and get involved with his talking cure?" Helen asks.

"We met only once, during the New Years holiday season of '27. I called on Freud who was staying in the Berlin home of once of his sons. Then, as I was, for better or worse, I guess, the living symbol of the physical sciences, at forty seven and Freud, at seventy, was my equal in the social sciences. I understood as much psychology as he did physics so we had a pleasant discussion. A friend later suggested that I undergo psycho-analysis but I could not agree to do so, as I should like very much in the dark and work to draw from that darkness of the unconscious, my the illumination of my theories. What do they say here, if it ain't broke don't try to fix it?"

Bohr adds, "Your fields of expertise were diverse, seemed almost irreconcilable but you might have had more in common than you might have imagined. Many years earlier in '07 and on, at the beginning of both of your investigations into what Hamlet called "the undiscovered country". You both were sensing a disconnect in the understandings of you respective fields of inquiry, and were confronting an obstacle which defeated everyone else exploring the problems. In both cases, the obstacle was a lack of more data, evidence. Yet rather than retreat from these challenges and look elsewhere, concede defeat and stop looking, you and Freud kept searching for that invisible universe."

Wheeler agrees. "I see that correlation as well, as Jung said between the wave function field and the psyche, and even the modern relativistic drama of Pirandello. It may very well be that art and science raw concepts and universal laws come from the same matrix of Universal thought. It may be that not a shining mechanism at the end of the Universe, but magic that awaits us, a Merlin waving his wand over the Cosmos..Researchers do that even now as we work in Los Alamos to find a new kind of implosion to accommodate the atomic weapons. Dick Feynman, the best young mind at Los Alamos is an example of that kind of intense thinking. Remember Niels when we went by his room at Los Alamos that night?"

Bohr laughs and takes some coffee.

"Yes, Dick is brilliant, on the cutting edge of physics and a character as well, with his bongo drums. He was one of the young guns at the place. We were walking by his room one night and heard this thrashing around sound. Thinking Dick was in trouble of some kind, we opened the door and found him rolling around on the floor as if physically trying to get into an atom's orbit and solve some problem. As you said it is the relentless drive to discover what is behind or below the surface, how the level of particle physics it superposition works, to make the invisible, visible. Correlation I think then is the threshold, when the paradox is broken, when two seemingly disparate ideas break into synthesis and a brand new paradigm emerges. The arts, psychology, sciences are coming together to join at the hips of each, a nexus wrought by the inner drive of genius to establish a universal law, as Albert said-- at the matrix of all creative thought. Pushing the twin frontiers of scientific research – the inner and the outer, as Freud did as well apparently, you all arrived at the impasse where thesis meets antithesis. Then you both spanned it, as Jung did as well with his impasse ultimately with Freud. Albert and Freud kept looking until you both discovered an entirely new kind of scientific evidence, evidence that no mere looking was going to reveal, evidence that lay beyond the realm of the visible."

Wheeler says in support of that thought. "The invisible has always been part of humanity's interactions with nature. Attempting to explain otherwise inexplicable phenomena, the ancients invented spirits, forms of gods; the Western World created God. Newton, when reaching the limits of his outer world, invoked gravity. Descartes reached the limits of his understanding of the inner universe, invoked the concept of consciousness."

Bohr adds, "Although Albert and Freud didn't initiate this second scientific revolution all by themselves, they did come to represent it and in large measure embody it. Their histories, stories of discovery reached unprecedented realms, relativity and the unconscious, and their further pursuits produced the creation of the two new sciences, cosmology and psychoanalysis. What these have allowed us to see, is not just going deeper than the years before into the disciplines but instead seeing it as a question of seeing – of perception itself, *how we see!* It is also a question of *thinking* about seeing , of conception, of how we think about what we see. What Albert and Freud have done shows us there is more to the Universe than we would ever find if all we did was look."

Margot comes in with more coffee and pours it for those who ask. "We can all move to the living room if you'd like. This is all very fascinating and it's good Helen is transcribing us as it could be the basis for a book. Having you all here like this is a real treat."

Everyone agrees to stay at the table. Adele asks Einstein, "Albert, when you got to Zurich how did your work go?"

"It was a joyful and productive time,"Einstein answers, "But I was tempted to apply for a full professorship at the German University in Prague and a salary increase so that we could afford a servant. As in Zurich, there were two applicants but I got it when Jaumann declined it in my favor. We moved to Prague though Mileva didn't want to go. In 1912, I got the chance to go back to Zurich to find Grossman who met me warmly at the Polytechnic and to whom I later went to when I was working around the clock to devise the field equations that would explain gravitation's role in the Universe or to know God's thoughts. Besso was there was there as well helping and dear Ehrenfest joined us for a spell. But soon after I returned to Zurich, Planck and Nernst arrived as a couple of corporate headhunters hoping to tempt me to join them at Berlin University with a bigger salary, membership in the exclusive Prussian Academy and freedom to pursue my work with no strings attached. So I went. When I told Planck that I was working on the gravitation problem, he warned me against it saying that if I succeeded, no one would believe me but if I did it and it proved true, I would be hailed as the next Copernicus. Mileva left Berlin for Zurich taking the boys with her which was the split-up that had been building. I wept when they left but it was inevitable that we separate. It was a mismatch from the start but time and circumstance had exacerbated it."

Margrethe asks, "How did you find Berlin?"

"I regarded most Germans as lunatics, especially the frenzied war advocates. But it was here, after eight years of trying, I was closing in on the true nature of the Universe and gravitation's role in it. I called a truce in the war of words with Mileva, focusing on our dear children and put everything on hold including Elsa, whom I'd been seeing, who lived close by with Margot and Ilse, both pre-teen then. Over the next five weeks in the fall of '15, I skipped meals, worked far into the night and I cooked everything together to same time."

Margot remembers. "I came by unexpectedly and found Papa boiling an egg in a saucepan of soup, intending to eat both!"

Einstein laughs. "But I had reviewed my work on special relativity and was not satisfied. There were two glaring holes in it. First relativity was based on inertial motions when in nature almost nothing is inertial. The second was more embarrassing: *it said nothing about gravity when gravity is everywhere!* I had to generalize relativity to include accelerations and gravity. I went back to my Patent Office days in Bern and the key insight I had had then, the thought picture that always gives the start to solution. I was slaving over patent applications and suddenly realized that if I had fallen backward in my chair I would be falling freely and the effect of gravity would be canceled by my acceleration, making me appear weightless. It was the happiest thought of my life. Energized, I kept working. Looking for, and finally finding the postulate behind general relativity...that the laws of physics in an accelerating frame are indistinguishable. Gravity could bend light. The largest gravitational field in the solar system is generated by the sun. If we could prove its gravity could warp space, bend light, it would be proved. I neared my goal with growing excitement, feeling as if something in me was about to snap, I got my answer!"

Still staring out to the night sky, as if that moment of his discovery, which some consider to be the supreme intellectual achievement of the human species was etched on the skin of the known universe and he was reading it back to himself, Einstein continues. "It was as if I were emerging from having seen, all these years, through a glass darkly and was finally seeing the cosmos face to face. Gravity, I concluded, is not a physical force of attraction acting through space as was commonly thought but a manifestation of the Universe's geometry. Space is curved or warped by the presence of matter and objects move through space along the shortest path following the contours of space. My theory explains the origin and destiny of the Universe, predicts that light passing a massive object will undergo a reddening and that a clock near a massive object will run slower than one at a distance from it, accounts for the inconsistent orbit of Mercury and points to the existence of gravity waves. The more massive the object, the greater the warping or curvature of space, the extent of which is greater near the object and grows less as the distance from it increases."

Wheeler says, "When I first heard of this brilliant concept which our friend Max Born considered a great work of art and the greatest feat of human thinking about nature, the most amazing combination of philosophical penetration, physical intuition and mathematical skill, I saw the beauty of it. Simply put, space tells matter how to move and matter tells space how to curve. Gravity is not a foreign and physical force acting through space – it's a manifestation of the geometry of space where mass is."

"Very well put, John," Bohr says. "One can only say what Albert did was akin to wizards work, as you said serious magic at the core of the cosmos. Where were the seeds that gave rise to this wonderfully unique structure? The mathematical framework had been prepared...Newton's theory, special relativity, Minkowski's idea of a four dimensional world, and Mach's powerful criticism of Newton's theory...But after that, what? Albert's intuitive feeling that the laws of physics of the time were somehow askew, the principle of equivalence, rhe principle of general covariance and why, essentially nothing else. By what magical clairvoyance did you choose just these principles to be your guide long before you knew where they would lead? That they should have led you to unique equations of so complex and yet simple of a sort, is in itself astounding."

Godel sits intensely interested in this and adds, "There are ten enormously complicated equations governing space-time curvature that I've been studying lately. If written out in full, instead of the compact tensor notation, they would fill a huge book with intricate symbols, in one form, millions of them. These field equations for general relativity there is something about them which is intensely beautiful and almost miraculous. Their power and utter naturalness in both form and content give them an indescribable beauty."

Einstein beams. "You all are to kind..Yet, dear Godel has let me know on one of our walks last week that he finds something in those gravitational equations that are...how did you put it Kurt?"

"Incomplete..but now offense meant dear friend."

"And none taken. I am most curious to see what you can do with time. And what philosophy you can derive from general relativity because as Merz has observed...the German man of science is a philosopher...But you all have failed to mention that I made one addition to my views of the universe that I would later regret. I held the conventional view that ours was a static Universe, that despite minor changes in the movements of stars and planets, the overall picture would remain the same

So to rebut the idea that our Universe is expanding which offended me, I invented and added what some scientists call a cosmological constant – often called the fudge factor – to bring it back to static. One of my most unforgivable mistakes."

Bohr adds, "Perhaps dear friend but down the line you may be somehow vindicated there as your prior predictions have been as your light rays being visible during an eclipse."

"I had concluded Niels, in 1911 that starlight passing close to the Sun should be slightly deflected toward it, following the curvature in space caused by its massive presence. Later I calculated that the deflection would be twice as great as I originally thought. Normally, these light rays would be invisible on earth obscured by the brightness of the Sun but they could be seen during an eclipse of the Sun – one of which was expected in 1916."

Bohr adds, "Which Eddington did with that solar eclipse, prompting the Nobel laureate J.J. Thompson of the Royal Society to solemnly say it was one of the greatest achievements in the history of modern thought, a whole continent of new scientific ideas...the greatest discovery in connection with gravitation since Newton's principles."

Einstein laughs. "Max Planck stayed up all night stayed up all night to see if the solar eclipse data would verify general relativity. If he had really understood relativity he would have gone to bed the way I did! The firestorm broke at that meeting of the Royal Society in November. I was suddenly transferred from a senior of physics in Berlin to a world figure, a worthy successor to, Lord, *Isaac Newton!* The philosopher Whitehead at that meting said that there was an atmosphere of intense interest that was exactly that of a Greek drama."

Bohr recalls. "I remember the way it burst out into the world. Eddington became, as Huxley had become Darwin's Bulldog, your champion using the full force of his scientific reputation and considerable debating skills to promote relativity."

Einstein returns to he table, sits quietly, then says, "My so called meteoric rise to scientific fame was due in part to the fact that the world was exhausted by the senseless savagery of World War 1 and was ready for the kind of mythic figure they made me but it gave me headaches. I dreamed I was burning in hell and the postman was the devil, eternally roaring at me, throwing new bundles of letters at my head because I had not answered the old ones. The world was a curious madhouse. I was in the center of cyclone of a relativity circus! There I was, stuck in the lunatic asylum called Berlin and wished I could move to Mars to observe the inmates through a telescope. But Berlin didn't have a monopoly on lunatics. I had just learned that the British had captured six miles of mud in the Battle of the Somme at a cost of almost a half a million men. When would such madness end? To have Hitler rise up out the ashes of the Weimar debacle and take Germany to even greater heights of insanity is incredible."

Wheeler drinks some coffee, gets up and walks to the end of the table. "With general relativity launched, along comes Heisenberg's Uncertainty and Niels' Complimentarity, the Copenhagen Interpretation of Quantum Mechanics. And those wonderful debates between Albert and Niels at the Six Solvay Conference which I consider the strangest yet the greatest in intellectual history that I know about. In thirty years I've never heard of a debate between two greater men, on a deeper issue with deeper consequences for understanding this strange world of ours. Margrethe, you were there with Niels at Solvay weren't you?"

"Yes indeed! Those debates were quite rigorous and vital and took Niels to the epitome of his friendly dual with Dr. Einstein. Niels, for Helen's record of this night can you describe how you and Heisenberg came up with the Copenhagen Interpretation?"

"Of course." Bohr says. "The recollection is fond as well as the memory of crossing swords with Albert at Solvay. Albert had, in '09 started it all by speaking of the dual nature of the electron, that it could be seen as both a wave and a particle depending how you measure it. De Broglie extended that to matter as well as matter waves."

"I like the audacity of his matter waves and supported it," Einstein says. But if matter was wave – like, what was the equation that it obeyed?"

Bohr responds. "Schrodinger did that for us in '25, grudgingly taking time from his holiday with his lady-love at Arosa and for the first time we had a detailed picture of the atom. Dirac picked up on this as well, boasted that all of chemistry could be explained as solutions of Schrodinger's equation, reducing chemistry to physics. He realized that Schrodinger's equation only applied to electrons moving at slow speeds compared to light and did not include relativity. So his electrons obeyed the full Einstein symmetry and could explain obscure properties of the electron, such as spin."

Wheeler interjects. "Then the paradox arises...if matter is a wave then precisely what is waving? A Schrodinger wave is like an ocean wave and eventually spreads out if left by itself. With enough time the wave function eventually dissipates over the entire Universe. But this violates everything we know about electrons. Subatomic particles were believed to be point-like objects that made definite jet-like streaks that could be measured on film. Though these wave functions had great success in describing the hydrogen atom, it did not seem possible it could describe an electron moving in free space. If Schrodinger's wave really represented the electron it would slowly dissipate and the Universe would dissolve."

Helen exclaims, "Heaven's that can't be good. What was done to break this new paradox?"

Einstein answers. "My good friend Max Born, in '26 broke the paradox but caused me concern. He said that Schrodinger's wave did not *actually* describe the electron but only the *probability* of finding the electron's location and speed, that the the motion of particles follows probability laws but probability itself propagates in conformity with the laws of of causality. In this new picture matter indeed consisted of point-like particles and not waves."

Wheeler grins. "I recall how disgusted Erwin was when he heard of Born's description, when he later called me. And later when you and Heisenberg merged Uncertainty and Complimentarity, He lamented that if he had only come up with the probable components of an electron he would not have bothered with it and not wasted the weekend away from his Dark Lady Muse of Arosa, whomever she may be. He went on poking fun at this view of his matter waves theory, by inventing the Schrodinger's Cat thought picture paradox."

"Which is?" Helen asks.

"Schrodinger "put" a cat in a closed container, with a mechanism that when the cat pushed the lever for food, there was a %50/%50 chance he would get food or get a shot of radioactive gas, killing it. He put the cat in the container in orbit around the world. In that state the cat is both alive and dead in the state of superposition, and you don't know until you look into the container whether the cat is alive or dead, In truth the cat can't be both alive and dead but it is the core of the paradox picture."

Bohr adds, "Werner took Born one step further. He and I agonized endlessly over puzzles of probability infesting this new theory. We went to Pauli for advice and became so obsessed with our work we almost had to be force fed. The weird, even uncanny world we'd discovered seemed to be the joint creation of an illusionist and the designer of a gambling casino. We had to get a solid and defensible scheme for our invention, to get it out of the realm of fantasy, or Albert and the other scientists at Solvay would tear us to shreds."

Margrethe chimes in. "Heisenberg confided in me how he had struggled so to come up Uncertainty and a way to make it feasible and a law of physics.

He was looking for that eureka moment that Dr.
Einstein had with relativity. After a frustrating
night with Niels of grappling unsuccessfully with
probabilities, he took a long walk down Faellle
Park, behind his university,constantly asking
himself how was it possible that he could not know
the precise location and speed of an electron, a
discrete object with mass albeit very small, as
described by Born. Then it hit him, everything
became clear. In order to know where an electron
was *you had to look at it!* That meant shining a light
beam at it. But the photons in the light beam would
collide with the electron making its position
uncertain. In other words, the act of observation
creates uncertainty."

Bohr continues. "Which gave us a new principle
of physics and a shock as well. The Uncertainty
Principle stated that one cannot determine both the
location and the velocity of the particle at the same
time. The more you know about one half of the
quantum world the less you know about the other
half hence. When we saw that this wasn't just a by-
product of the crudeness of our instruments, that it
probably described a law of nature, as Kurt said, it
was pushed into a kind of mystique and religion.
What upset Albert about the theory it presumed
that not even God could determine the location of
an electron. We were stunned. But instead of

feeling elated as Albert was with his discovery of relativity, we were in despair, made repeated attempts to disprove ourselves, sought repudiation everywhere."

Wheeler commiserates. "But Albert, have not you been mistakenly cast as the last dinosaur of classical physics? You were never totally dismissive of quantum theory were you? Didn't you say at Solvay although you tried hard to refute it with Niels that quantum theory demands a certain respect especially on a practical level, applied to industry and electronics...you even recommended Schrodinger and Heisenberg share in the '26 Nobel Prize...aren't you saying to be complete it had to merge, hand in glove, with general relativity, to give the full picture of Unification as it was in the first millisecond after the Big Bang?"

Einstein nods. "Yes, that is my position now. After Solvay I shifted my strategy in talking of quantum mechanics, not to refute it outright as it was, as I felt in my mind, not the true Jacob, but say that it was incomplete without hooking up with general relativity. I shifted context of the argument, to a larger stage, in the search for Unification, the theory of everything as one writer put it. The Mind of the Old One as I said can be "read" to some degree, but I was sure that this random idea of the dynamics of quantum mechanics was not the way The Old One. This view of Heisenberg and Niels when it came out I thought presumptively limited His ability to be super omniscient in every aspect of the physical world, both micro and macro. For my part at least I am convinced that He doesn't play dice with the Universe. If had known that the quantum revolution I helped start would introduce chance into physics, I would have become a cobbler!"

Margrethe says with a light laugh, "I remember when Dr. Einstein made that remark at Solvay about God not playing dice with the Universe, Niels told him he shouldn't tell God what to do! Niels, after Heisenberg's breakthrough onto Uncertainty, how did you arrive at Complimentarity? I recall you were quite frustrated there for a time."

"Yes, I was. Heisenberg and I understood Albert's resistance to our strange results knowing your conviction and philosophical attitude, that the world could be completely divided into objective and subjective spheres and he hypothesis that one should be able to make precise statements about the objective side. You and Werner met and talked about that in Berlin in '25, did you not Albert?"

"Yes, he came to the university in the Spring of '25 to lecture on your joint work. I invited him to my home and told him that what he had said had sounded extremely strange. He assumed the existence of electrons inside the atom and he was probably right to do so but he refused to consider their orbits even though we can observe electron tracks of moisture in a cloud chamber. I asked him did he seriously believe that none but the observable magnitudes must go into a physical theory."

Bohr recalls. "He told me later that he met you and was surprised you'd said that, reminded him that he had done the same with relativity. He pointed out that you did stress the fact that it is impermissible to speak of absolute time cannot be observed, that only, that only clock readings sre relevant to the determination of time."

Einstein recalls. "I remember that, yes,and admitted that I may have used the same reasoning back then with relativity but added that it was nonsense, just the same. In reality, the very opposite happens. It is the theory which decides what we observe, When I said God doesn't throw dice, I was offended by the consummate arrogance of the man of science to say that because he couldn't see the electron, both its position and speed simultaneously that God is also locked in that darkness. Meister Eckert, the Christian mystic said the eye with which I see God he sees me."

"Werner and I were in too much of a rush to test the theory and at first didn't concern ourselves with the moving our work into metaphysics. But, as has been mentioned, it's taken on a life on its own, purporting to say that somehow quantum theory is a mystical state or has arisen from the Chinese concept of the Tao, when it is merely way to measure sub atomic particles. Schrodinger tried to trivialize it, make it absurd with his cat-in-the-box thought puzzle, that the cat can be both dead and alive in the same quantum moment. Even that absurd picture is gaining a measure of truth in some experimental circles. But it stumped me as well as Werner:how could quantum mechanics be so weird yet get such excellent experimental results?"

Margrethe adds this. "It just got too much for Niels. So in mid-argument with Heisenberg, he went on a skiing vacation in Norway, hoping to clear his mind and come up with something to break the paradox that had arisen demanding reconciliation and Niels you returned with..."

"*Complimentarity!* I returned to Copenhagen full of energy born of an intuitive insight that burst into my brain as I flew down the slopes, then bound up the stairs at the Institute. My answer was to retreat from the Newtonian subject/object language the either/or mindset. I agreed with Heisenberg that the quest for an unambiguous definition of the atom had to be abandoned. I resolved the dilemma by making the best of it, that knowledge of these aspects when separated would give the most accurate picture as possible on the subatomic world. If you wanted to see the electron as a wave you measured it with one kind of instrument and if a particle then another instrument is used and then the pictures compliment each other for a true picture of the electron."

Adele asks, "So then, just as Dr. Einstein found his special relativity in the paradox between Newton and Maxwell, your Complimentarity came out of the paradox that the electron can be both a wave and a particle?"

"Precisely Adele! Instead of regarding the dual nature of the wave/particle nature of the electron as another problem, the breaking of the Newtonian Paradox, free of that paralyzing reductionism, by way of a sort of quantum nirvana illuminated by my Complimentarity. However, contrasting such phenomena may first appear it must be information about the atomic object which can be expressed in common language without ambiguity. What we came up with, what Werner suggested we call the Copenhagen Interpretation of Quantum Mechanics, was that were two truths rather than one alone, that the two together offer science and man a more complete view and understanding of the atomic world than either could do separately and thus in the end a clearer view of the visible world built out of invisible substratum of the atom. Instead of division, I showed the parts, the divisions separate, even contradictory aspect compliments the other when the viewer changes her perception, angle of observation and expectation."

Margrethe adds, "Adele , it's like that photo that is used in psychology classes to show the power of observation and perception that creates a personality. Perceived one way, it is a picture of a beautiful girl. A change in the seeing of it, the girl shifts into the image of a crone. *You* decide what you want to see once you are introduced to this way of thinking in the quantum mechanics manner."

Bohr smiles at his wife's analogy. "It's not a world reduced to either/or. When I set out this bridge over Uncertainty, to a new kind of uncertainty to my students who were like novice Zen priests sitting with a Zen Master trying to solve a koan with logic and Newtonian terms, like trying to "solve" the Zen koan, we know the sound of two hands clapping, *show me*, the sound of one had clapping – inferring s picture of the solution. They for most art understood Complimentarity still stuck in he Newtonian lock-box, rather than a synthesis. I kept them thinking, pressing what Dick Feynman said, that if you keep asking about how could quantum theory could be so weird, it will drive you crazy. I quoted them Abbe of Galina, "One cannot know the front of somebody without showing the back of someone else."

Wheeler says, "Your quantum theory postulates not only uncertainty and duality but also the ghostly behavior of matter that is created out of nothing,then just mysteriously vanishes."

Bohr Adds, "But on the positive side it explains why atoms and molecules keep their identity, their shapes and their patterns in spite of collisions and perturbations, why gold is gold wherever we find it and in the last instance, why the same flowers bloom every spring."

Einstein says, "I was largely unaware of what you and Heisenberg were up to Niels, the intensity of your efforts to clinch your schizophrenic picture of the nature of atomic structure. I missed the September Solvay Conference near Italy's Lake Como which I refused to attend because the conference was sponsored by Mussolini's Fascist Government."

"With you absent, Albert, we worked hard, as you would recall John, to fine tune our theory. We encouraged everyone to raise every possible objection to Uncertainty and Complimentarity, then countered them, we thought, very well. But the big question was would Albert be won over at the Sixth Solvay Conference in Belgium? We were confident we could since our work was experimentally verified. And after all Abert was considered by most as the grandfather of quantum duality. Little did we suspect that Albert would come out swinging as it were against it as a law of nature, even though we knew what he had told Werner, that he had rejected it years ago."

Margrethe takes some water and adds, "Niels and I arrived there in October. He was quite confident, after the dry run with Dr. Einstein absent, there were few, if any, questions he could not answer."

Wheeler rises and walks to the window and speaks of that historic meeting. "Lorentz, the conference chairman, who, sadly only had a few months to live, asked Niels at the start of the session, to spell out the problems facing quantum theory. As you know Niels, with his soft voice and often indistinct pronunciation, shifting between dialects...no offense dear friend...it was not the easiest presentation to follow, as you rushed on, crowding the blackboard with diagrams and equations."

Margrethe recalls dramatically. "Then Dr.

Einstein rose and the room grew silent as he gave a low key – key but emphatic rejection of quantum theory as a poor attempt to destroy determinism. He characterized Niels and Heisenberg, who sat worried as he spoke and their supporters, though good naturally and with a wry smile as walking on eggs to avoid physical reality. Before he sat down, several scientists stood and shouted in various languages for permission to speak as Dr. Lorentz pounded his gravel like a frustrated judge. Dr. Ehrenfest walked to the blackboard and wrote, 'The Lord did there confound the language of all the earth!' "

There's a light laughter around the table then Bohr says, "It wasn't clear whether he meant the babble of voices, my indistinct, subdued delivery or Uncertainty itself but it brought a roar of laughter and provided a lighthearted conclusion to the first days discussion."

Einstein agrees. "Yes, that was amusing. Later, near the end of the conference my friend Ehrenfest chided me for aggressive nature of my attack on quantum theory, was ashamed of me arguing thus against it just as my opponents argued against relativity theory. But even his friendly admonition went unheard until later when as I said I shifted strategies regarding my view of quantum theory, from out right hostility to the better view that gave quantum theory its due but only as a method of viewing and describing the action on the level of particle physics."

Bohr relates. "That night Heisenberg told me how it was again driven home to him, on hearing Albert's brilliant and firm refutation, how terribly difficult it was for you to give up on an attitude on which your entire scientific approach and career have been based. You were not prepared, Werner suggested, letting us do that which amounted to pulling the ground from under your feet. And at that next session you came up with an ingenious proof I assume you had worked out in advance, that really gave me a fit."

Einstein glances at the open window as a falling star arcs across the cosmos. "Then, I was quite determined to disprove it to the world, and tried mightily to destroy the demon of Uncertainty. My weapon of choice was the thought picture of a box containing radiation with a hole with a shutter, which opens briefly, releasing a photon. Measuring the weight before and after the photon's release, he box weighs less. Because of the equivalence of matter and energy and the time of the opening of the box, we can know how much energy the box contains and the time of opening without any uncertainty. Hence Uncertainty is wrong."

Bohr emits a low groan as if reliving the feeling of panic when he first heard this sublime refutation from the greatest physicist of our time. "To me this was a heavy blow. At first hearing I had saw no solution. I remained unhappy all through the evening, going from person to person trying to persuade them all that this could not happen, that if Albert were right this would mean the end of physics. But I could think of no refutation."

Wheeler recalls as well, with awe. "The two of them, Albert and Niels...watching them with fascination and awe so fiercely engaged in this wonderful debate was thrilling was thrilling. I'll never forget the sight of those two opponents and the weight of what the outcome could be, how it would affect our humanity, leaving the university club, Albert a majestic figure walking calmly with a faintly ironic smile, and Niels..."

Bohr laughs at himself. "And I trotting alongside, extremely upset. When I talked to Ehrenfest later that evening all I could say was, 'Einstein, Einstein, Einstein.' But after an intense, restive night, in a moment of deep sleep, in a dream sort of as Kulke "saw" the benzene ring, serpent biting its tail, I used your own relativity to refute your proof!"

Wheeler describes Bohr's reasoning. "You

presented the defense that next morning. You pointed out the because the box weighed less than before it would rise slightly in earth's gravity, But according to relativity tije speeds up as gravity gets weaker. Any uncertainty in measuring time of the shutter would be translated into an uncertainty in measuring the box's position which will result in an uncertainty in the weight which is in turn reflected in the energy and momentum of the box. When everything is put together the agree precisely with the Uncertainty Principle. Niels had successfully defended Uncertainty!"

Margot enters with fresh coffee, offers it around the table. Maja is sleeping nicely after we read an exiting section of The Odyssey. Anyone for more coffee? I hope I haven't missed much."

Einstein holds up his cup and Margot pours some coffee and passes it around. "We were recalling Dr.Bohr's and my "debate" at the Sixth Solvay. To my dismay but respecting his great brilliance, Niels refuted my first salvo, then through the entire conference as I worked and up and threw more attacks on Uncertainty. Niels rebutted them all! After the conference, we went to Ehrenfest's home and continued out friendly skirmish. Later that night, I walked with de Broglie. Louis was a leading proponent og quantum theory and a fine mind. Homeward bound from Belgium, I went to

Paris with him, left him at the train station with a wave and an encouraging. "Carry on. You are on the right track!" I recommended him for '29 Nobel. The conference was wonderfully stimulating but very tiring for me."

Margot puts down the coffee pot. "Papa came home. Elsa found him tired and unusually subdued. When an invitation came from Millikan to visit Cal Tech, he declined and admitted he was completely exhausted."

Einstein stirs sugar in his coffee, stares at the spoon for a moment. "I have thought a hundred times as much about quantum problems than I have about general relativity. When you think it out, you must conclude that the process of observation determines the final state state of the object which almost make one laugh at the patent absurdity of it, thinking in Newtonian language. If that be so, for the ultimate reality in nature why not posit an Old Ones eye, a cosmic consciousness can *see it!* I've heard it said in a metaphysical mode, that our one great story or world myth, the shape of man's ontogic essence and existential heft is that we all come forth from one ground of being, mani-festations of time on a timeless ground , a kind of shadow field. And we play the game in the shadow field, enacting our side of the polarity with all our might suspecting that our Self is on the other side of ourselves if we could only see from the position of the middle. Like Plato's cave, watching the shadows on the wall that we think are humanity. Before Newton we didn't know that we were only shadows on a wall of reality, but now with this new quantum theory at least we are aware of that seeming anomaly. Artists and scientists propose ways to take us to the position of the middle, and stall out usually but the Eye of the Old One, can see it all. When a tree falls in a rain forest does it make a sound if there is no one there to hear it? Newton, myself, Shrodinger with his cat-in-the-box paradox would say certainly it does. But you Niels even now would say that even before it is *seen* by the human eye, it exists in potentia where a tree can exist in all possible states...fallen...upright, sapling, burnt, rotten..."

Helen laughs. "When the subject is broached by visitors, Dr. Einstein would ask, "Does the Moon exist because a mouse looks at it?"

Bohr responds. "Albert, after Sixth Solvay and after regaining your fervor of the mind, you *still* came after Werner and I with your students Podolsky and Rosen, and you EPR experiment. How did that emerge?"

"With the EPR experiment we came up with

what we felt was he definitive critique of quantum theory. Suppose that an atom emits two electrons in opposite directions, each spinning like a top, pointing either up or down, in opposite ways so the total spin is zero even though you don't know what the spins are. If you wait long enough these electrons could be billions of miles apart but if you find out which way one is spinning even though they are light years away, the other one will spin in the opposite direction. This means that a measurement in one part of the Universe instantly determines the state of an electron on the other side of the Universe, seemingly in violation of my special relativity. I called it 'spooky-at-a-distance'. I ultimately disliked EPR as if meant that the Universe was non-local."

Godel looks away in a moment of reflection. "The philosophical implications of EPR are rather startling. It means that some atoms in our body may be connected with an invisible web of atoms on the other side of the Universe, atoms perhaps in an alien body in some parallel world, an alien which may have viruses we cannot defeat if they invade out world, perhaps arriving through a worm hole that Albert and Rosen concocted with trheir 'Einstein-Rosen Bridge" in '35."

FERVOR OF THE MINDS/Nagel/80

Bohr recalls. "When we heard of Albert's EPR experiment we dropped everything and went to work, trying to clear up such a misunderstanding at once. Using my students as sounding boards, we came up with a rejoinder which I dictated to Margrethe at once, then had Werner read it and we published."

Margrethe recalls. "I recall the relief in them both when the paper came out. They had withstood the challenge but at a price. They has to concede to Dr. Einstein that the quantum mechanical Universe was indeed non local. Everything in the Universe is meshed together in a dense network a cosmic entanglement. We are attached to nature as a giant web and what we do to one strand of the web, vibrates and resonates throughout the web even to the end of the cosmos."

Einstein explains. "Out EPR may have not demolished quantum theory as we hoped it would but it exceeded in proving that it, already pretty bizarre, got even more bizarre with the non local rule, revealed how really crazy quantum theory was. But EPR gas over the years been misunderstood with scores of speculations that one could build an EPR faster than light radio or we can use it to send signals back in time or use it to affect telepathy between people and worlds."

Wheeler remarks, "But EPR didn't negate relativity and Albert I think had the last laugh after EPR was examined experimentally. No information can be transmitted faster than the speed of light via the RPR method. We have a mathematician on campus who always wore a pink sock and a green sock. If you know one foot had the green sock, you knew the other sock was pink. Yet no signal went from one foot to another. In other words knowing something is entirely different from sending that knowledge hence a world of difference between possession of information and its transmission."

Wheeler walks to the end of the table, turns to the group. "So to recapitulate now we have two towering branches of physics, relativity and quantum theory. The sum total of all human knowledge about the physical Universe could be summarized by these two theories , with relativity giving us the structure of the macro world of the very large objects, the theory of the big bang and black holes. And quantum theory deals with the micro world, the subatomic. The Copenhagen view sets up a wall between to tow, the hard copy world and he world of phantoms. Didn't Heisenberg say once that atoms are not things, that they are only tendencies? I call that level of quantum superposition a smokey dragon. Macro and micro will never meet."

Einstein takes issue with that ultimatum. "Never say never John. We are only at the edge of the Zeitgeist and I predict that the new particle accelerators, the atom smashers we are developing as we speak will smash through that wall render part of the micro world as part of the macro and an other Nobel will fly off the wall they way they did in the late twenties when the young lions in our craft were willing to apply quantum theory and I was forced to admit that quantum mechanics is the most successful theory of our period. As I said, after my attempts to discredit it helped to solidify its import, and am working with it to created Unification with general relativity. I was happy to recommend that Heisenberg share the '29 Nobel with Schrodinger. My goal was nothing short of cosmic in scope, to swallow up quantum theory in its entirety into my Unification. I am sure that during the first millisecond after the big bang when the Universe was expanding much faster than the speed of light, when matter and anti matter were dueling and matter won by a billion of a percent of energy, that that was enough to keep the Universe going to what it is today. In a way you can say that the Universe in all its forms and shapes are really leftovers from the effort clash. At that moment quantum theory and general relativity were one."

Wheeler goes on. "And your work again with Rosen came up with a novel way to sidestep quantum theory as a theory of first, fundamental principles, to have the electron emerge naturally as a consequence of your theory."

"Yes John, it made sense to us then to do that. In most theories elementary particles emerge as singularities, that is, regions where equations blow up, creating singularities. We wanted to derive quantum theory from a deeper realm and reasoned that we needed a theory totally free of singularities, like the solitons which resemble kinks in space, smooth not singular, can bounce off each other and still maintain their shape. We used two Schwarz-child black holes, using two parallel sheets where you could cut out each black hole and glue the two sheets back together. This obtaining a smooth singularity free solution. I thought solitons might represent a subatomic particle. Thus quantum particles can be viewed as tiny black holes. Our EPR Bridge could be viewed as wormholes, shortcuts through space and time like a gateway or portal that connects two parallel sheets of paper."

Godel seems entranced by the concept of worm holes between parts of the Universe, rises and goes to the window ans stares out. "Wormholes, connecting two Universes..but why only two one might boldly ask! We are introduced to wormholes in Alice In Wonderland, when Alice goes down he rabbit hole, and stepping into Kafka's quantum shape shifting worlds in The Castle, where Alice and K are in worlds where logic and reality are stood on their heads, with the Mad Hatter and the absurd self contradictory, self canceling beaurocracy with the same kind of insanity that gives the idiots at the FBI power to keep the genius our dear Dr. Einstein, from working on the weapon that could very well save us from Nazi annihilation." He smiles at his dear friend. "And what would do to time Herr Professor?"

Einstein regards his young friend with soft eyes: his face crinkles up into a matrix of amusement somewhat as it had on occasion on their walks when Godel had said something really wirklich verruki – really cracked. "Kurt, you love your fables don't you? But I too seek a higher beauty, an artistic paradigm in my equations for mathematical patterns as those of a painter or poet must be beautiful like colors or words must fit it all together in a harmonious way. Beauty is the first test. There is no place for ugly mathematics or physics. They must be elegant in their application to the physical world. But I fear anything really innovative and world changing is only invented in youth. Later one becomes more experienced - and more stupid."

www.ingramcontent.com/pod-product-compliance
Lightning Source LLC
Chambersburg PA
CBHW070911180526
45168CB00005B/2003